The Techno-Optimist Manifesto

Marc Andreessen
The Techno-Optimist Manifesto

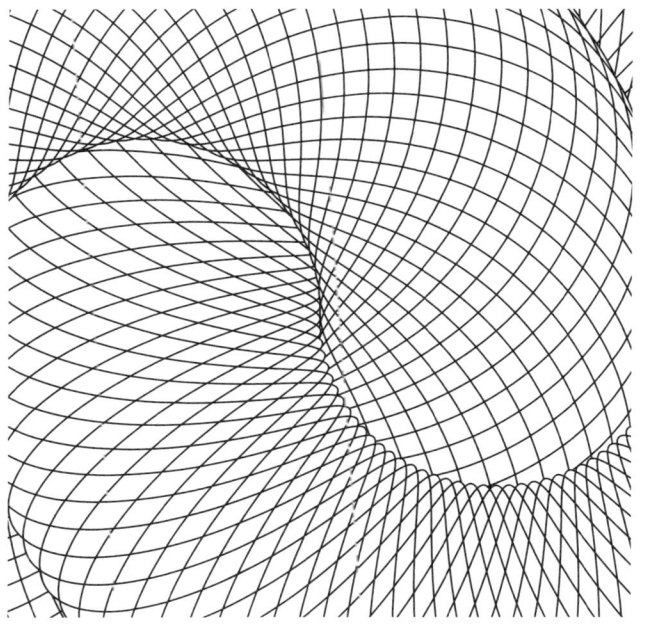

NETWORK PRESS

LIMITED FOUNDER'S EDITION
ISBN: 979-8-9916415-0-0

TRADE PAPERBACK
ISBN: 979-8-9916415-1-7

Copyright © 2024 by Marc Andreessen

Cover and interior design by Shaun Roberts

All rights reserved. No portion of this book may be used or reproduced in any manner whatsoever without written permission from the publisher and copyright holder.

Printed in the United States of America

For information, contact:
support@network.press

NETWORK PRESS
www.network.press

Contents

1.
Why Software Is Eating the World

15.
It's Time to Build

27.
The Techno-Optimist Manifesto

LIES — 34

TRUTH — 36

TECHNOLOGY — 38

MARKETS — 42

THE TECHNO-CAPITAL MACHINE — 48

INTELLIGENCE — 52

ENERGY — 56

ABUNDANCE — 60

NOT UTOPIA, BUT CLOSE ENOUGH — 64

BECOMING TECHNOLOGICAL SUPERMEN — 66

TECHNOLOGICAL VALUES — 70

THE MEANING OF LIFE — 76

THE ENEMY — 80

THE FUTURE — 86

PATRON SAINTS OF TECHNO-OPTIMISM — 88

HTTPS://A16Z.COM/WHY-SOFTWARE-IS-EATING-THE-WORLD/

POSTED AUGUST 20, 2011[*]

---

[*] ORIGINALLY PUBLISHED IN THE *WALL STREET JOURNAL* ON AUGUST 20, 2011.

Software is eating the world

More than ten years after the peak of the 1990s dot-com bubble, a dozen or so new Internet companies such as Facebook and Twitter are sparking controversy in Silicon Valley, due to their rapidly growing private market valuations, and even the occasional successful IPO. With scars from the heyday of Webvan and Pets.com still fresh in the investor psyche, people are asking, "Isn't this just a dangerous new bubble?"

I, along with others, have been arguing the other side of the case. (I am co-founder and general partner of venture capital firm Andreessen-Horowitz, which has invested in Facebook, Groupon, Skype, Twitter, Zynga, and Foursquare, among others. I am also personally an investor in LinkedIn.) We believe that many of the prominent new Internet companies are building real, high-growth, high-margin, highly defensible businesses.

Today's stock market actually hates technology, as shown by all-time low price/earnings ratios for major public technology companies. Apple, for example, has a P/E ratio of around 15.2—about the same as the broader stock market, despite Apple's immense profitability and dominant market position (Apple in the last couple weeks became the biggest company in America, judged by market capitalization, surpassing ExxonMobil).

And, perhaps most telling, you can't have a bubble when people are constantly screaming "Bubble!"

But too much of the debate is still around financial valuation, as opposed to the underlying intrinsic value of the best of Silicon Valley's new companies. My own theory is that we are in the middle of a dramatic and broad technological and economic shift in which software companies are poised to take over large swaths of the economy.

More and more major businesses and industries are being run on software and delivered as online services—from movies to agriculture to national defense. Many of the winners are Silicon Valley–style entrepreneurial technology companies that are invading and overturning established industry structures. Over the next ten years, I expect many more industries to be disrupted by software, with new world-beating Silicon Valley companies doing the disruption in more cases than not.

Why is this happening now?

Six decades into the computer revolution, four decades since the invention of the microprocessor, and two decades into the rise of the modern Internet, all of the technology required to transform industries through software finally works and can be widely delivered at global scale.

Over two billion people now use the broadband Internet, up from perhaps fifty million a decade ago, when I was at Netscape, the company I co-founded. In the next ten years, I expect at least five billion people worldwide to own smartphones, giving every individual with such a phone instant access to the full power of the Internet, every moment of every day.

On the back end, software programming tools and Internet-based services make it easy to launch new global software-powered start-ups in many industries—without the need to invest in new infrastructure and train new employees. In 2000, when my partner Ben Horowitz was CEO of the first cloud computing company, Loudcloud, the cost of a customer running a basic Internet application was approximately $150,000 per month. Running that same application today in Amazon's cloud costs about $1,500 per month.

With lower start-up costs and a vastly expanded market for online services, the result is a global economy that for the first time will be fully digitally wired—the dream of every cyber-visionary of the early 1990s, finally delivered, a full generation later.

Perhaps the single most dramatic example of this phenomenon of software eating a traditional business is the suicide of Borders and the corresponding rise of Amazon.

In 2001, Borders agreed to hand over its online business to Amazon under the theory that online book sales were non-strategic and unimportant.

Oops

Today, the world's largest bookseller, Amazon, is a software company—its core capability is its amazing software engine for selling virtually everything online, no retail stores necessary. On top of that, while Borders was thrashing in the throes of impending bankruptcy, Amazon rearranged its website to promote its Kindle digital books over physical books for the first time. Now even the books themselves are software.

Today's largest video service by number of subscribers is a software company: Netflix. How Netflix eviscerated Blockbuster is an old story, but now other traditional entertainment providers are facing the same threat. Comcast, Time Warner, and others are responding by transforming themselves into software companies with efforts such as TV Everywhere, which liberates content from the physical cable and connects it to smartphones and tablets.

Today's dominant music companies are software companies, too: Apple's iTunes, Spotify, and Pandora.

Traditional record labels increasingly exist only to provide those software companies with content. Industry revenue from digital channels totaled $4.6 billion in 2010, growing to 29% of total revenue from 2% in 2004.

Today's fastest-growing entertainment companies are videogame makers—again, software—with the industry growing to $60 billion from $30 billion five years ago. And the fastest-growing major videogame company is Zynga (maker of games including FarmVille), which delivers its games entirely online. Zynga's first-quarter revenues grew to $235 million this year, more than double the revenues from a year earlier. Rovio, maker of Angry Birds, is expected to clear $100 million in revenue this year (the company was nearly bankrupt when it debuted the popular game on the iPhone in late 2009). Meanwhile, traditional videogame powerhouses such as Electronic Arts and Nintendo have seen revenues stagnate and fall.

The best new movie production company in many decades, Pixar, was a software company. Disney—Disney!—had to buy Pixar, a software company, to remain relevant in animated movies.

Photography, of course, was eaten by software long ago. It's virtually impossible to buy a mobile phone that doesn't include a software-powered camera, and photos are uploaded automatically to the Internet for permanent archiving and global sharing. Companies such as Shutterfly, Snapfish, and Flickr have stepped into Kodak's place.

Today's largest direct-marketing platform is a software company—Google. Now it's been joined by Groupon, Living Social, Foursquare, and others, which are using software to eat the retail marketing industry. Groupon generated over $700 million in revenue in 2010, after being in business for only two years.

Today's fastest-growing telecom company is Skype, a software company that was just bought by Microsoft for $8.5 billion. CenturyLink, the third-largest telecom company in the U.S., with a $20 billion market cap, had 15 million access lines at the end of June 30—declining at an annual rate of about 7%. Excluding the revenue from its Qwest acquisition, CenturyLink's revenue from these legacy services declined by more than 11%. Meanwhile, the two biggest telecom companies, AT&T and Verizon, have survived by transforming themselves into software companies, partnering with Apple and other smartphone makers.

LinkedIn is today's fastest-growing recruiting company. For the first time ever, on LinkedIn, employees can maintain their own resumes for recruiters to search in real time— giving LinkedIn the opportunity to eat the lucrative $400 billion recruiting industry.

Software is also eating much of the value chain of industries that are widely viewed as primarily existing in the physical world. In today's cars, software runs the engines, controls safety features, entertains passengers, guides drivers to destinations, and connects each car to mobile, satellite, and GPS networks. The days when a car aficionado could repair his or her own car are long past, due primarily to the high software content. The trend toward hybrid and electric vehicles will only accelerate the software shift—electric cars are completely computer controlled. And the creation of software-powered driverless cars is already underway at Google and the major car companies.

Today's leading real-world retailer, Wal-Mart, uses software to power its logistics and distribution capabilities, which it has used to crush its competition. Likewise for FedEx, which is best thought of as a software network that happens to have trucks, planes, and distribution hubs attached. And the success or failure of airlines today and in the future hinges on their ability to price tickets and optimize routes and yields correctly—with software.

Oil and gas companies were early innovators in supercomputing and data visualization and analysis, which are crucial to today's oil and gas exploration efforts. Agriculture is increasingly powered by software as well, including satellite analysis of soils linked to per-acre seed selection software algorithms.

The financial services industry has been visibly transformed by software over the last thirty years. Practically every financial transaction, from someone buying a cup of coffee to someone trading a trillion dollars of credit default derivatives, is done in software. And many of the leading innovators in financial services are software companies, such as Square, which allows anyone to accept credit card payments with a mobile phone, and PayPal, which generated more than $1 billion in revenue in the second quarter of this year, up 31% over the previous year.

Health care and education, in my view, are next up for fundamental software-based transformation. My venture capital firm is backing aggressive start-ups in both of these gigantic and critical industries. We believe both of these industries, which historically have been highly resistant to entrepreneurial change, are primed for tipping by great new software-centric entrepreneurs.

Even national defense is increasingly software-based. The modern combat soldier is embedded in a web of software that provides intelligence, communications, logistics, and weapons guidance. Software-powered drones launch air strikes without putting human pilots at risk. Intelligence agencies do large-scale data mining with software to uncover and track potential terrorist plots.

Companies in every industry need to assume that a software revolution is coming. This includes even industries that are software-based today. Great incumbent software companies such as Oracle and Microsoft are increasingly threatened with irrelevance by new software offerings like Salesforce.com and Android (especially in a world where Google owns a major handset maker).

In some industries, particularly those with a heavy real-world component such as oil and gas, the software revolution is primarily an opportunity for incumbents. But in many industries, new software ideas will result in the rise of new Silicon Valley–style start-ups that invade existing industries with impunity. Over the next ten years, the battles between incumbents and software-powered insurgents will be epic. Joseph Schumpeter, the economist who coined the term "creative destruction," would be proud.

And while people watching the values of their 401(k) plans bounce up and down the last few weeks might doubt it, this is a profoundly positive story for the American economy, in particular. It's not an accident that many of the biggest recent technology companies—including Google, Amazon, eBay, and more—are American companies. Our combination of great research universities, a pro-risk business culture, deep pools of innovation-seeking equity capital, and reliable business and contract law is unprecedented and unparalleled in the world.

Still, we face several challenges

First of all, every new company today is being built in the face of massive economic headwinds, making the challenge far greater than it was in the relatively benign '90s. The good news about building a company during times like this is that the companies that do succeed are going to be extremely strong and resilient. And when the economy finally stabilizes, look out—the best of the new companies will grow even faster.

Secondly, many people in the U.S. and around the world lack the education and skills required to participate in the great new companies coming out of the software revolution. This is a tragedy since every company I work with is absolutely starved for talent. Qualified software engineers, managers, marketers, and salespeople in Silicon Valley can rack up dozens of high-paying, high-upside job offers any time they want, while national unemployment and underemployment is sky-high. This problem is even worse than it looks because many workers in existing industries will be stranded on the wrong side of software-based disruption and may never be able to work in their fields again. There's no way through this problem other than education, and we have a long way to go.

Finally, the new companies need to prove their worth. They need to build strong cultures, delight their customers, establish their own competitive advantages, and yes, justify their rising valuations. No one should expect building a new high-growth, software-powered company in an established industry to be easy. It's brutally difficult.

I'm privileged to work with some of the best of the new breed of software companies, and I can tell you they're really good at what they do. If they perform to my and others' expectations, they are going to be highly valuable cornerstone companies in the global economy, eating markets far larger than the technology industry has historically been able to pursue.

Instead of constantly questioning their valuations, let's seek to understand how the new generation of technology companies are doing what they do, what the broader consequences are for businesses and the economy, and what we can collectively do to expand the number of innovative new software companies created in the U.S. and around the world.

That's the big opportunity. I know where I'm putting my money.

HTTPS:// A16Z.COM/ ITS-TIME-TO- BUILD/

POSTED APRIL 18, 2020

Every Western institution was unprepared for the coronavirus pandemic, despite many prior warnings. This monumental failure of institutional effectiveness will reverberate for the rest of the decade, but it's not too early to ask why, and what we need to do about it.

Many of us would like to pin the cause on one political party or another, on one government or another. But the harsh reality is that it all failed—no Western country, or state, or city was prepared—and despite hard work and often extraordinary sacrifice by many people within these institutions. So the problem runs deeper than your favorite political opponent or your home nation.

Part of the problem is clearly foresight, a failure of imagination. But the other part of the problem is what we didn't *do* in advance, and what we're failing to do now. And that is a failure of action, and specifically our widespread inability to *build*.

We see this today with the things we urgently need but don't have. We don't have enough coronavirus tests, or test materials—including, amazingly, cotton swabs and common reagents. We don't have enough ventilators, negative pressure rooms, and ICU beds. And we don't have enough surgical masks, eye shields, and medical gowns—as I write this, New York City has put out a desperate call for rain ponchos to be used as medical gowns. Rain ponchos! In 2020! In America!

We also don't have therapies or a vaccine—despite, again, years of advance warning about bat-borne coronaviruses. Our scientists will hopefully invent therapies and a vaccine, but then we may not have the manufacturing factories required to scale their production. And even then, we'll see if we can deploy therapies or a vaccine fast enough to matter—it took scientists five years to get regulatory testing approval for the new Ebola vaccine after that scourge's 2014 outbreak, at the cost of many lives.

In the U.S., we don't even have the ability to get federal bailout money to the people and businesses that need it. Tens of millions of laid-off workers and their families, and many millions of small businesses, are in serious trouble *right now*, and we have no direct method to transfer them money without potentially disastrous delays. A government that collects money from all its citizens and businesses each year has never built a system to distribute money to us when it's needed most.

Why do we not have these things? Medical equipment and financial conduits involve no rocket science whatsoever. At least therapies and vaccines are hard! Making masks and transferring money are not hard. We could have these things but we chose not to—specifically we chose not to have the mechanisms, the factories, the systems to make these things. We chose not to ***build***.

You don't just see this smug complacency, this satisfaction with the status quo and the unwillingness to build, in the pandemic, or in health care generally. You see it throughout Western life, and specifically throughout American life.

You see it in housing and the physical footprint of our cities. We can't build nearly enough housing in our cities with surging economic potential—which results in crazily skyrocketing housing prices in places like San Francisco, making it nearly impossible for regular people to move in and take the jobs of the future. We also can't build the cities themselves anymore. When the producers of HBO's *Westworld* wanted to portray the American city of the future, they didn't film in Seattle or Los Angeles or Austin—they went to Singapore. We should have gleaming skyscrapers and spectacular living environments in all our best cities at levels way beyond what we have now; where are they?

You see it in education. We have top-end universities, yes, but with the capacity to teach only a microscopic percentage of the 4 million new 18 year olds in the U.S. each year, or the 120 million new 18 year olds in the world each year. Why not educate every 18 year old? Isn't that the most important thing we can possibly do? Why not build a far larger number of universities, or scale the ones we have way up? The last major innovation in K-12 education was Montessori, which traces back to the 1960s; we've been doing education research that's never reached practical deployment for fifty years since; why not build a lot more great K-12 schools using everything we now know? We know one-to-one tutoring can reliably increase education outcomes by two standard deviations (the Bloom two-sigma effect); we have the Internet; why haven't we built systems to match every young learner with an older tutor to dramatically improve student success?

You see it in manufacturing. Contrary to conventional wisdom, American manufacturing output is higher than ever, but why has so much manufacturing been offshored to places with cheaper manual labor? We know how to build highly automated factories. We know the enormous number of higher paying jobs we would create to design and build and operate those factories. We know—and we're experiencing right now!—the strategic problem of relying on offshore manufacturing of key goods. Why aren't we building Elon Musk's "alien dreadnoughts"—giant, gleaming, state of the art factories producing every conceivable kind of product, at the highest possible quality and lowest possible cost—all throughout our country?

You see it in transportation. Where are the supersonic aircraft? Where are the millions of delivery drones? Where are the high-speed trains, the soaring monorails, the hyperloops, and yes, the flying cars?

Is the problem money? That seems hard to believe when we have the money to wage endless wars in the Middle East and repeatedly bail out incumbent banks, airlines, and carmakers. The federal government just passed a $2 trillion coronavirus rescue package in two weeks! Is the problem capitalism? I'm with Nicholas Stern when he says that capitalism is how we take care of people we don't know—all of these fields are highly lucrative already and should be prime stomping grounds for capitalist investment, good both for the investor and the customers who are served. Is the problem technical competence? Clearly not, or we wouldn't have the homes and skyscrapers, schools and hospitals, cars and trains, computers and smartphones, that we already have.

The problem is desire. We need to *want* these things. The problem is inertia. We need to want these things more than we want to prevent these things. The problem is regulatory capture. We need to want new companies to build these things, even if incumbents don't like it, even if only to force the incumbents to build these things. And the problem is will. We need to build these things.

And we need to separate the imperative to build these things from ideology and politics. Both sides need to contribute to building.

The right starts out in a more natural, albeit compromised, place. The right is generally pro-production, but is too often corrupted by forces that hold back market-based competition and the building of things. The right must fight hard against crony capitalism, regulatory capture, ossified oligopolies, risk-inducing offshoring, and investor-friendly buybacks in lieu of customer-friendly (and, over a longer period of time, even more investor-friendly) innovation.

It's time for full-throated, unapologetic, uncompromised political support from the right for aggressive investment in new products, in new industries, in new factories, in new science, in big leaps forward.

The left starts out with a stronger bias toward the public sector in many of these areas. To which I say, prove the superior model! Demonstrate that the public sector can build better hospitals, better schools, better transportation, better cities, better housing. Stop trying to protect the old, the entrenched, the irrelevant; commit the public sector fully to the future. Milton Friedman once said the great public sector mistake is to judge policies and programs by their intentions rather than their results.

Instead of taking that as an insult, take it as a challenge—build new things and show the results!

Show that new models of public-sector health care can be inexpensive and effective—how about starting with the VA? When the next coronavirus comes along, blow us away! Even private universities like Harvard are lavished with public funding; why can't 100,000 or 1 million students a year attend Harvard? Why shouldn't regulators and taxpayers demand that Harvard build? Solve the climate crisis by building—energy experts say that all carbon-based electrical power generation on the planet could be replaced by a few thousand new zero-emission nuclear reactors, so let's build those. Maybe we can start with ten new reactors? Then one hundred? Then the rest?

In fact, I think building is how we reboot the American dream. The things we build in huge quantities, like computers and TVs, drop rapidly in price. The things we don't, like housing, schools, and hospitals, skyrocket in price. What's the American dream? The opportunity to have a home of your own, and a family you can provide for. We need to break the rapidly escalating price curves for housing, education, and health care, to make sure that every American can realize the dream, and the only way to do that is to build.

Building isn't easy, or we'd already be doing all this. We need to demand more of our political leaders, of our CEOs, our entrepreneurs, our investors. We need to demand more of our culture, of our society. And we need to demand more from one another. We're all necessary, and we can all contribute, to building.

Every step of the way, to everyone around us, we should be asking the question, What are you building? What are you building directly, or helping other people to build, or teaching other people to build, or how are you taking care of people who are building? If the work you're doing isn't either leading to something being built or taking care of people directly, we've failed you, and we need to get you into a position, an occupation, a career where you can contribute to building. There are always outstanding people in even the most broken systems—we need to get all the talent we can on the biggest problems we have, and on building the answers to those problems.

I expect this essay to be the target of criticism. Here's a modest proposal to my critics. Instead of attacking my ideas of what to build, conceive your own! What do you think we should build? There's an excellent chance I'll agree with you.

Our nation and our civilization were built on production, on building. Our forefathers and foremothers built roads and trains, farms and factories, then the computer, the microchip, the smartphone, and uncounted thousands of other things that we now take for granted, that are all around us, that define our lives and provide for our well-being. There is only one way to honor their legacy and to create the future we want for our own children and grandchildren, and that is to build.

HTTPS:// A16Z.COM/ THE-TECH- NO-OPTI- MIST-MANI- FESTO/

POSTED OCTOBER 16, 2023

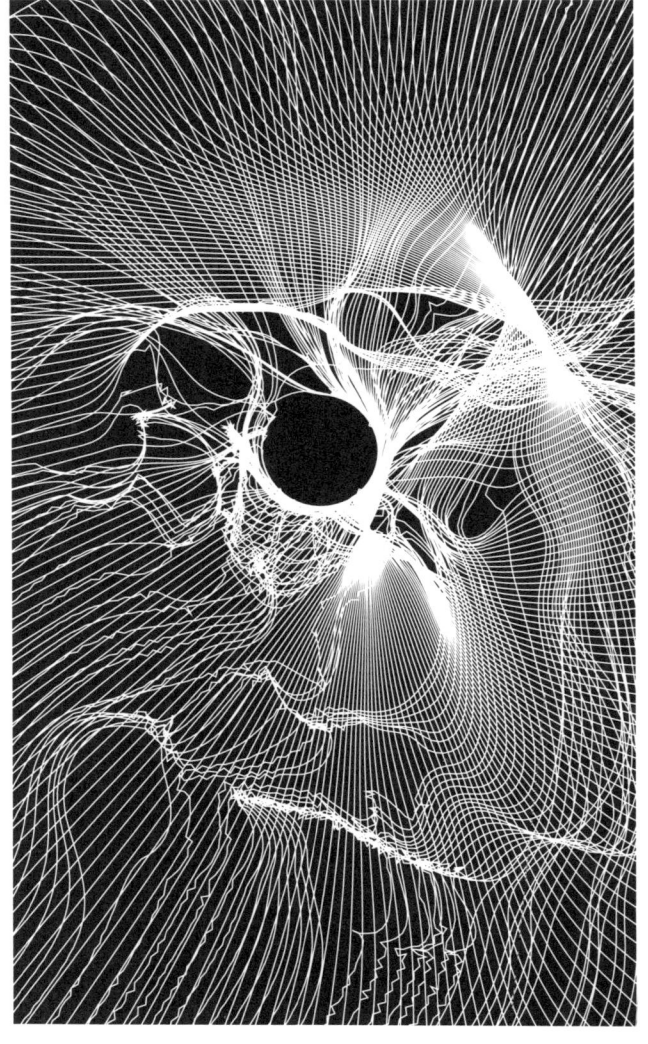

> You live in a deranged age—
> more deranged than usual,
> because despite great scientific
> and technological advances,
> man has not the faintest idea of
> who he is or what he is doing.
>
> —WALKER PERCY

Our species is 300,000 years old. For the first 290,000 years, we were foragers, subsisting in a way that's still observable among the Bushmen of the Kalahari and the Sentinelese of the Andaman Islands. Even after Homo sapiens embraced agriculture, progress was painfully slow. A person born in Sumer in 4000 B.C. would find the resources, work, and technology available in England at the time of the Norman Conquest or in the Aztec Empire at the time of Columbus quite familiar. Then, beginning in the 18th century, many people's standard of living skyrocketed. What brought about this dramatic improvement, and why?

—**MARIAN TUPY**

> There's a way to do it better. Find it.
>
> —THOMAS EDISON

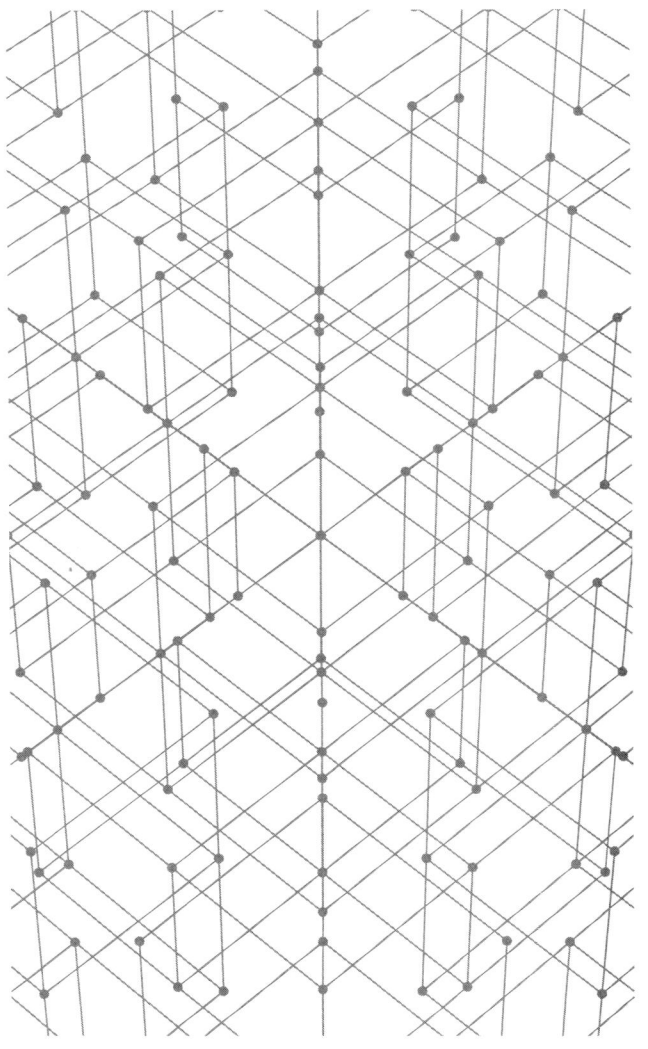

Lies

We are being lied to.

We are told that technology takes our jobs, reduces our wages, increases inequality, threatens our health, ruins the environment, degrades our society, corrupts our children, impairs our humanity, threatens our future, and is ever on the verge of ruining everything.

We are told to be angry, bitter, and resentful about technology.

We are told to be pessimistic.

The myth of Prometheus—in various updated forms like Frankenstein, Oppenheimer, and Terminator—haunts our nightmares.

We are told to denounce our birthright—our intelligence, our control over nature, our ability to build a better world.

We are told to be miserable about the future.

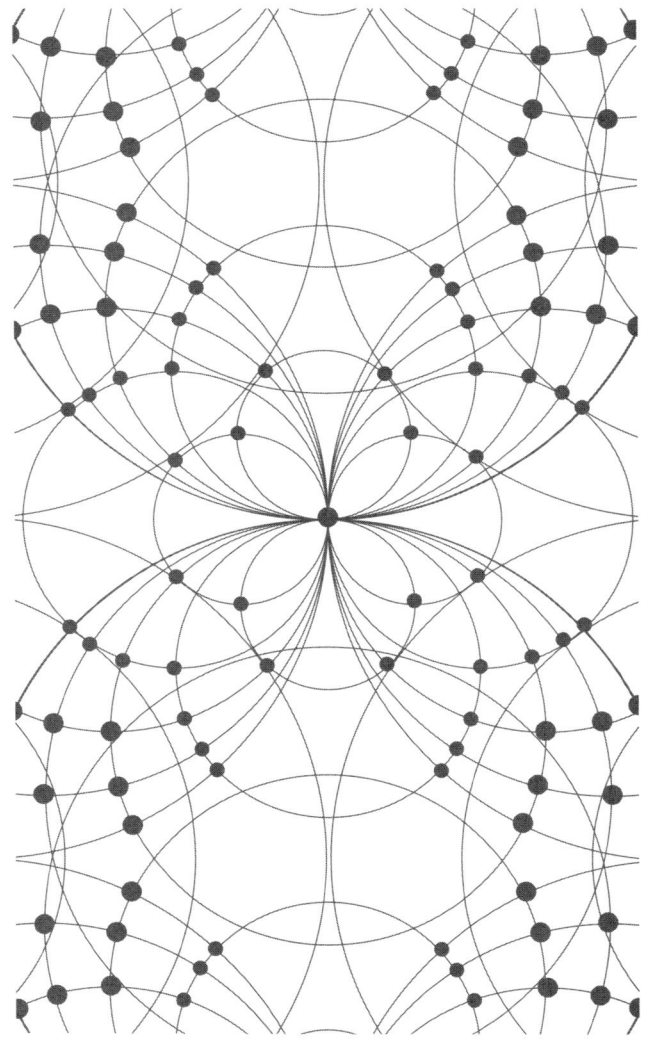

Truth

Our civilization was built on technology.

Our civilization is built on technology.

Technology is the glory of human ambition and achievement, the spearhead of progress, and the realization of our potential.

For hundreds of years, we properly glorified this—until recently.

I am here to bring the good news.

We can advance to a far superior way of living, and of being.

We have the tools, the systems, the ideas.

We have the will.

It is time, once again, to raise the technology flag.

It is time to be Techno-Optimists.

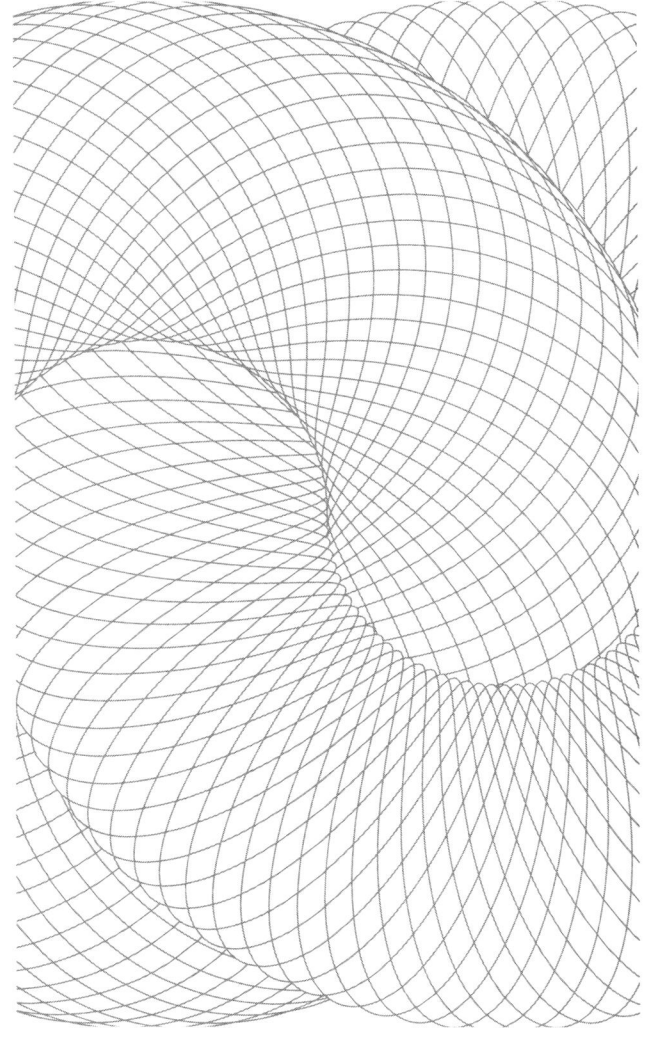

Technology

Techno-Optimists believe that societies, like sharks, grow or die.

We believe growth is progress—leading to vitality, expansion of life, increasing knowledge, higher well-being.

We agree with Paul Collier when he says, "Economic growth is not a cure-all, but lack of growth is a kill-all."

We believe everything good is downstream of growth.

We believe not growing is stagnation, which leads to zero-sum thinking, internal fighting, degradation, collapse, and ultimately death.

There are only three sources of growth: population growth, natural resource utilization, and technology.

Developed societies are depopulating all over the world, across cultures—the total human population may already be shrinking.

Natural resource utilization has sharp limits, both real and political.

And so the only perpetual source of growth is technology.

In fact, technology—new knowledge, new tools, what the Greeks called *techne*—has always been the main source of growth, and perhaps the only cause of growth, as technology made both population growth and natural resource utilization possible.

We believe technology is a lever on the world—the way to make more with less.

Economists measure technological progress as productivity growth: How much more we can produce each year with fewer inputs, fewer raw materials. Productivity growth, powered by technology, is the main driver of economic growth, wage growth, and the creation of new industries and new jobs, as people and capital are continuously freed to do more important, valuable things than in the past. Productivity growth causes prices to fall, supply to rise, and demand to expand, improving the material well-being of the entire population.

We believe this is the story of the material development of our civilization; this is why we are not still living in mud huts, eking out a meager survival and waiting for nature to kill us.

We believe this is why our descendents will live in the stars.

We believe that there is no material problem—whether created by nature or by technology—that cannot be solved with more technology.

> *We had a problem of starvation,*
> so we invented the Green Revolution.
>
> *We had a problem of darkness,*
> so we invented electric lighting.
>
> *We had a problem of cold,*
> so we invented indoor heating.
>
> *We had a problem of heat,*
> so we invented air-conditioning.
>
> *We had a problem of isolation,*
> so we invented the Internet.
>
> *We had a problem of pandemics,*
> so we invented vaccines.
>
> *We have a problem of poverty,*
> so we invent technology to create abundance.

Give us a real world problem, and we can invent technology that will solve it.

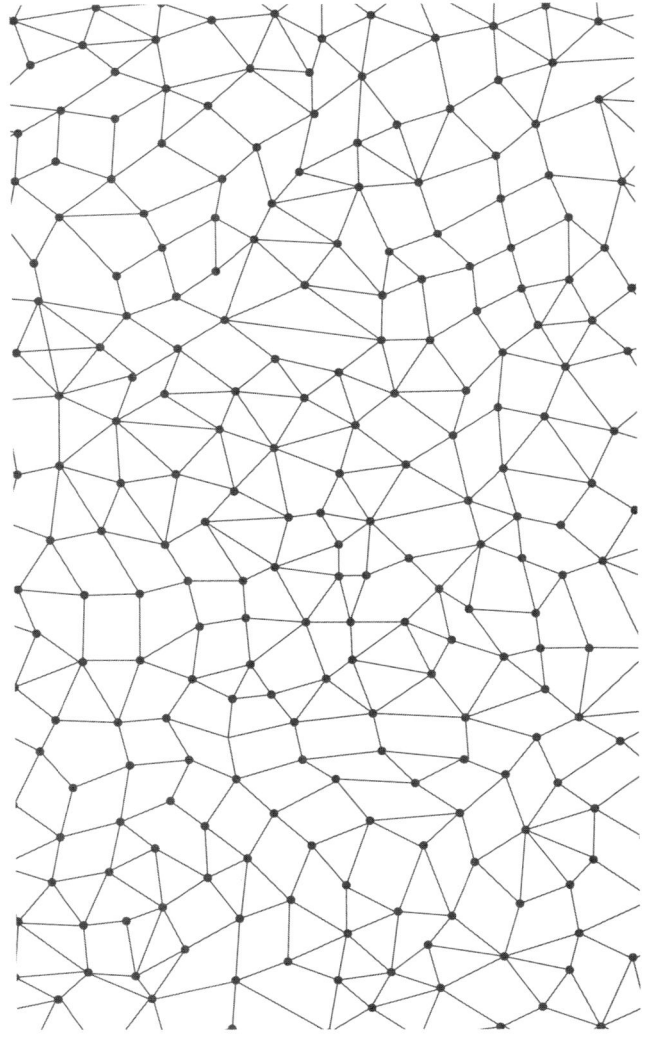

Markets

We believe free markets are the most effective way to organize a technological economy. Willing buyer meets willing seller, a price is struck, both sides benefit from the exchange or it doesn't happen. Profits are the incentive for producing supply that fulfills demand. Prices encode information about supply and demand. Markets cause entrepreneurs to seek out high prices as a signal of opportunity to create new wealth by driving those prices down.

We believe the market economy is a discovery machine, a form of intelligence—an exploratory, evolutionary, adaptive system.

We believe Hayek's Knowledge Problem overwhelms any centralized economic system. All actual information is on the edges, in the hands of the people closest to the buyer. The center, abstracted away from both the buyer and the seller, knows nothing. Centralized planning is doomed to fail, the system of production and consumption is too complex. Decentralization harnesses complexity for the benefit of everyone; centralization will starve you to death.

We believe in market discipline. The market naturally disciplines—the seller either learns and changes when the buyer fails to show, or exits the market. When market discipline is absent, there is no limit to how crazy things can get. The motto of every monopoly and cartel, every centralized institution not subject to market discipline is: "We don't care, because we don't have to." Markets prevent monopolies and cartels.

We believe markets lift people out of poverty—in fact, markets are by far the most effective way to lift vast numbers of people out of poverty, and always have been. Even in totalitarian regimes, an incremental lifting of the repressive boot off the throat of the people and their ability to produce and trade leads to rapidly rising incomes and standards of living. Lift the boot a little more, even better. Take the boot off entirely, who knows how rich everyone can get.

We believe markets are an inherently individualistic way to achieve superior collective outcomes.

We believe markets do not require people to be perfect, or even well-intentioned—which is good, because, have you met people? As Adam Smith notes:

> It is not from the benevolence of the butcher, the brewer, or the baker that we expect our dinner, but from their regard to their own self-interest. We address ourselves not to their humanity but to their self-love, and never talk to them of our own necessities, but of their advantages.

David Friedman points out that people only do things for other people for three reasons—love, money, or force. Love doesn't scale, so the economy can only run on money or force. The force experiment has been run and found wanting. Let's stick with money.

We believe the ultimate moral defense of markets is that they divert people who otherwise would raise armies and start religions into peacefully productive pursuits.

We believe markets, to quote Nicholas Stern, are how we take care of people we don't know.

We believe markets are the way to generate societal wealth for everything else we want to pay for, including basic research, social welfare programs, and national defense.

We believe there is no conflict between capitalist profits and a social welfare system that protects the vulnerable. In fact, they are aligned—the production of markets creates the economic wealth that pays for everything else we want as a society.

We believe central economic planning elevates the worst of us and drags everyone down; markets exploit the best of us to benefit all of us.

We believe central planning is a doom loop; markets are an upward spiral.

The economist William Nordhaus has shown that creators of technology are only able to capture about 2% of the economic value created by that technology. The other 98% flows through to society in the form of what economists call social surplus. Technological innovation in a market system is inherently philanthropic, by a 50:1 ratio. Who gets more value from a new technology, the single company that makes it, or the millions or billions of people who use it to improve their lives? QED.

We believe in David Ricardo's concept of comparative advantage—as distinct from competitive advantage, comparative advantage holds that even someone who is best in the world at doing everything will buy most things from other people, due to opportunity cost. Comparative advantage in the context of a properly free market guarantees high employment regardless of the level of technology.

We believe a market sets wages as a function of the marginal productivity of the worker. Therefore technology—which raises productivity—drives wages up, not down. This is perhaps the most counterintuitive idea in all of economics, but it's true, and we have three hundred years of history that prove it.

We believe in Milton Friedman's observation that human wants and needs are infinite.

We believe markets also increase societal well-being by generating work in which people can productively engage. We believe a Universal Basic Income would turn people into zoo animals to be farmed by the state. Man was not meant to be farmed; man was meant to be useful, to be productive, to be proud.

We believe technological change, far from reducing the need for human work, increases it, by broadening the scope of what humans can productively do.

We believe that since human wants and needs are infinite, economic demand is infinite, and job growth can continue forever.

We believe markets are generative, not exploitative; positive sum, not zero sum. Participants in markets build on one another's work and output. James Carse describes finite games and infinite games—finite games have an end, when one person wins and another person loses; infinite games never end, as players collaborate to discover what's possible in the game. Markets are the ultimate infinite game.

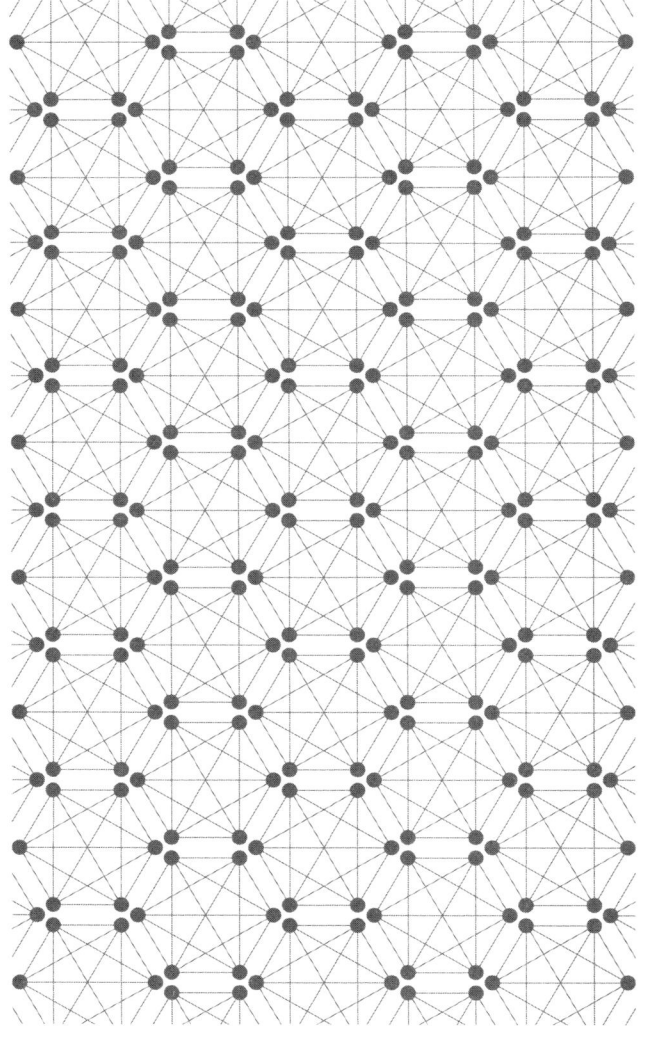

The Techno-Capital Machine

Combine technology and markets and you get what Nick Land has termed the techno-capital machine, the engine of perpetual material creation, growth, and abundance.

We believe the techno-capital machine of markets and innovation never ends, but instead spirals continuously upward. Comparative advantage increases specialization and trade. Prices fall, freeing up purchasing power, creating demand. Falling prices benefit everyone who buys goods and services, which is to say everyone. Human wants and needs are endless, and entrepreneurs continuously create new goods and services to satisfy those wants and needs, deploying unlimited numbers of people and machines in the process. This upward spiral has been running for hundreds of years, despite continuous howling from Communists and Luddites.

Indeed, as of 2019, before the temporary COVID disruption, the result was the largest number of jobs at the highest wages and the highest levels of material living standards in the history of the planet.

The techno-capital machine makes natural selection work for us in the realm of ideas. The best and most productive ideas win, and are combined and generate even better ideas. Those ideas materialize in the real world as technologically enabled goods and services that never would have emerged de novo.

Ray Kurzweil defines his Law of Accelerating Returns: Technological advances tend to feed on themselves, increasing the rate of further advance.

We believe in accelerationism—the conscious and deliberate propulsion of technological development—to ensure the fulfillment of the Law of Accelerating Returns. To ensure the techno-capital upward spiral continues forever.

We believe the techno-capital machine is not anti-human—in fact, it may be the most pro-human thing there is. It serves us. The techno-capital machine works for us. All the machines work for us.

We believe the cornerstone resources of the techno-capital upward spiral are intelligence and energy—ideas, and the power to make them real.

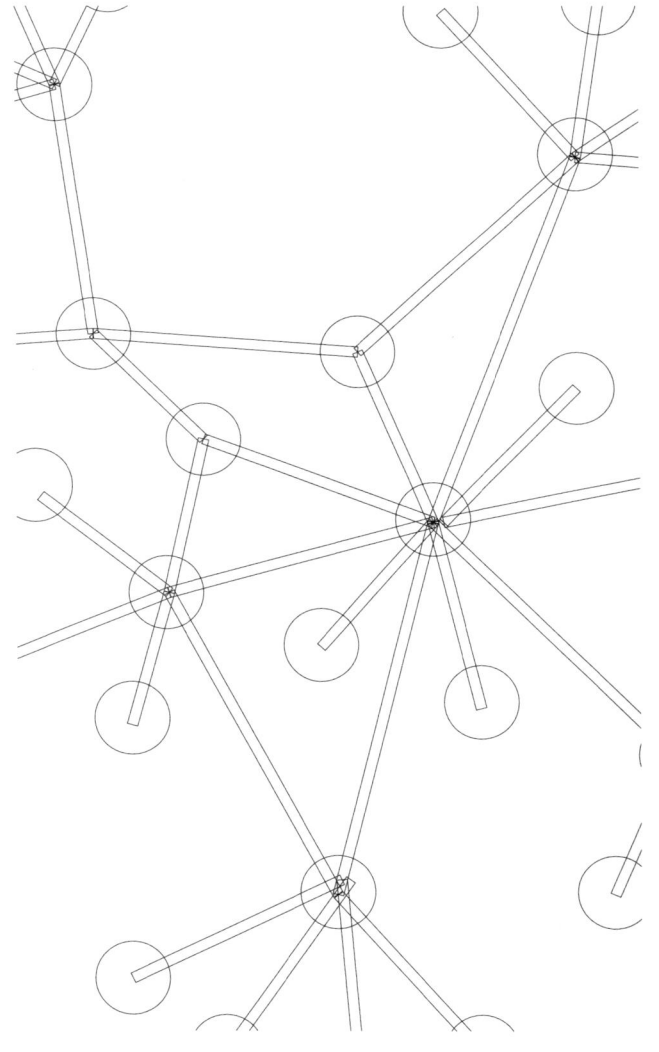

Intelligence

We believe intelligence is the ultimate engine of progress. Intelligence makes everything better. Smart people and smart societies outperform less smart ones on virtually every metric we can measure. Intelligence is the birthright of humanity; we should expand it as fully and broadly as we possibly can.

We believe intelligence is in an upward spiral—first, as more smart people around the world are recruited into the techno-capital machine; second, as people form symbiotic relationships with machines into new cybernetic systems such as companies and networks; third, as Artificial Intelligence ramps up the capabilities of our machines and ourselves.

We believe we are poised for an intelligence takeoff that will expand our capabilities to unimagined heights.

We believe Artificial Intelligence is our alchemy, our Philosopher's Stone—***we are literally making sand think.***

We believe Artificial Intelligence is best thought of as a universal problem solver. And we have a lot of problems to solve.

We believe Artificial Intelligence can save lives—if we let it. Medicine, among many other fields, is in the Stone Age compared to what we can achieve with joined human and machine intelligence working on new cures. There are scores of common causes of death that can be fixed with AI, from car crashes to pandemics to wartime friendly fire.

We believe any deceleration of AI will cost lives. Deaths that were preventable by the AI that was prevented from existing is a form of murder.

We believe in Augmented Intelligence just as much as we believe in Artificial Intelligence. Intelligent machines augment intelligent humans, driving a geometric expansion of what humans can do.

We believe Augmented Intelligence drives marginal productivity, which drives wage growth, which drives demand, which drives the creation of new supply ... with no upper bound.

Energy

Energy is life. We take it for granted, but without it, we have darkness, starvation, and pain. With it, we have light, safety, and warmth.

We believe energy should be in an upward spiral. Energy is the foundational engine of our civilization. The more energy we have, the more people we can have, and the better everyone's lives can be. We should raise everyone to the energy consumption level we have, then increase our energy 1,000x, then raise everyone else's energy 1,000x as well.

The current gap in per-capita energy use between the smaller developed world and larger developing world is enormous. That gap will close—either by massively expanding energy production, making everyone better off, or by massively reducing energy production, making everyone worse off.

We believe energy need not expand to the detriment of the natural environment. We have the silver bullet for virtually unlimited zero-emissions energy today—nuclear fission. In 1973, President Richard Nixon called for Project Independence, the construction of one thousand nuclear power plants by the year 2000, to achieve complete U.S. energy independence. Nixon was right; we didn't build the plants then, but we can now, anytime we decide we want to.

Atomic Energy Commissioner Thomas Murray said in 1953:

> For years the splitting atom, packaged in weapons, has been our main shield against the barbarians. Now, in addition, it is a God-given instrument to do the constructive work of mankind.

Murray was right too.

We believe a second energy silver bullet is coming—nuclear fusion. We should build that as well. The same bad ideas that effectively outlawed fission are going to try to outlaw fusion. We should not let them.

We believe there is no inherent conflict between the techno-capital machine and the natural environment. Per-capita U.S. carbon emissions are lower now than they were one hundred years ago, even without nuclear power.

We believe technology is the solution to environmental degradation and crisis. A technologically advanced society improves the natural environment; a technologically stagnant society ruins it. If you want to see environmental devastation, visit a former Communist country. The socialist U.S.S.R. was far worse for the natural environment than the capitalist U.S. Google the Aral Sea.

We believe a technologically stagnant society has limited energy at the cost of environmental ruin; a technologically advanced society has unlimited clean energy for everyone.

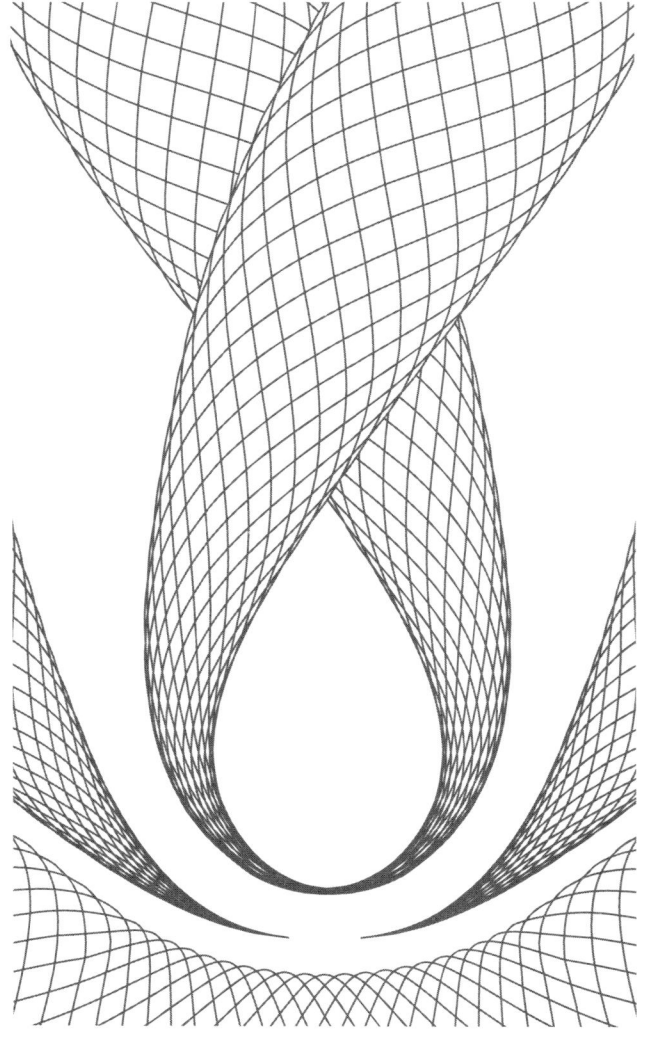

Abundance

We believe we should place intelligence and energy in a positive feedback loop, and drive them both to infinity.

We believe we should use the feedback loop of intelligence and energy to make everything we want and need abundant.

We believe the measure of abundance is falling prices. Every time a price falls, the universe of people who buy it get a raise in buying power, which is the same as a raise in income. If a lot of goods and services drop in price, the result is an upward explosion of buying power, real income, and quality of life.

We believe that if we make both intelligence and energy "too cheap to meter," the ultimate result will be that all physical goods become as cheap as pencils. Pencils are actually quite technologically complex and difficult to manufacture, and yet nobody gets mad if you borrow a pencil and fail to return it. We should make the same true of all physical goods.

We believe we should push to drop prices across the economy through the application of technology until as many prices are effectively zero as possible, driving income levels and quality of life into the stratosphere.

We believe Andy Warhol was right when he said,

> What's great about this country is America started the tradition where the richest consumers buy essentially the same things as the poorest. You can be watching TV and see Coca-Cola, and you can know that the President drinks Coke, Liz Taylor drinks Coke, and just think, you can drink Coke, too. A Coke is a Coke and no amount of money can get you a better Coke than the one the bum on the corner is drinking. All the Cokes are the same and all the Cokes are good.

Same for the browser, the smartphone, the chatbot.

We believe that technology ultimately drives the world to what Buckminster Fuller called "ephemeralization"—what economists call "dematerialization."

Fuller:

> Technology lets you do more and more with less and less until eventually you can do everything with nothing.

We believe technological progress therefore leads to material abundance for everyone.

We believe the ultimate payoff from technological abundance can be a massive expansion in what Julian Simon called "the ultimate resource"—people.

We believe, as Simon did, that people are the ultimate resource—with more people come more creativity, more new ideas, and more technological progress.

We believe material abundance therefore ultimately means more people—a lot more people—which in turn leads to more abundance.

We believe our planet is dramatically underpopulated, compared to the population we could have with abundant intelligence, energy, and material goods.

We believe the global population can quite easily expand to fifty billion people or more, and then far beyond that as we ultimately settle other planets.

We believe that out of all of these people will come scientists, technologists, artists, and visionaries beyond our wildest dreams.

We believe the ultimate mission of technology is to advance life both on Earth and in the stars.

Not Utopia, But Close Enough

However, we are not Utopians.

We are adherents to what Thomas Sowell calls the Constrained Vision.

We believe the Constrained Vision—contra the Unconstrained Vision of Utopia, Communism, and Expertise—means taking people as they are, testing ideas empirically, and liberating people to make their own choices.

We believe in not Utopia, but also not Apocalypse.

We believe change only happens on the margin—but a lot of change across a very large margin can lead to big outcomes.

While not Utopian, *we believe* in what Brad DeLong terms "slouching toward Utopia"—doing the best that fallen humanity can do, making things better as we go.

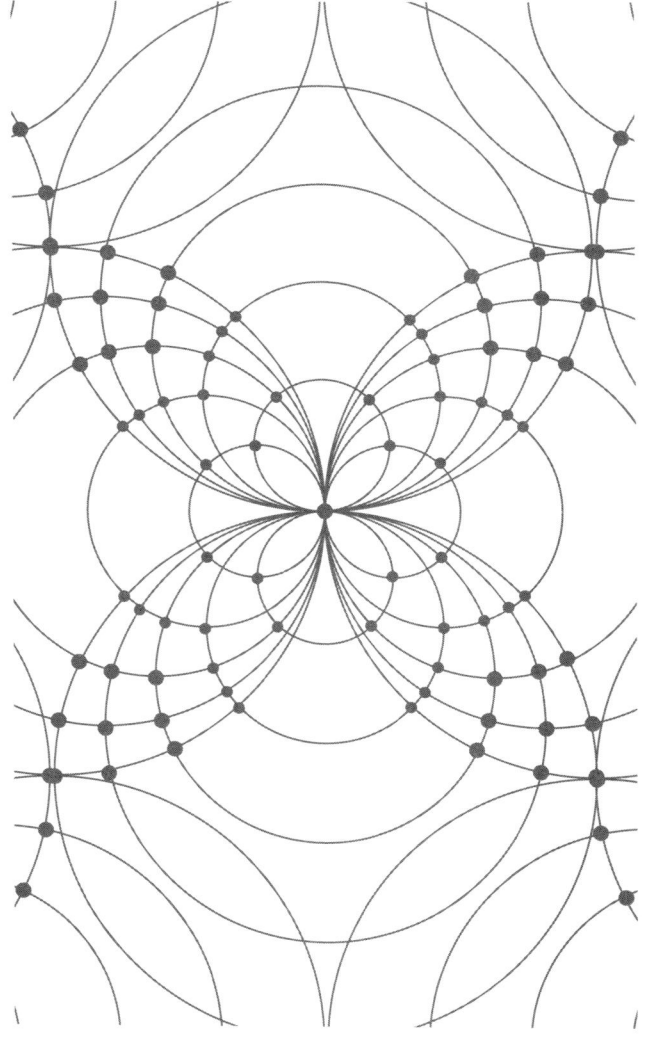

Becoming Technological Supermen

We believe that advancing technology is one of the most virtuous things that we can do.

We believe in deliberately and systematically transforming ourselves into the kind of people who can advance technology.

We believe this certainly means technical education, but it also means going hands-on, gaining practical skills, working within and leading teams—aspiring to build something greater than oneself, aspiring to work with others to build something greater as a group.

We believe the natural human drive to make things, to gain territory, to explore the unknown can be channeled productively into building technology.

We believe that while the physical frontier, at least here on Earth, is closed, the technological frontier is wide open.

We believe in exploring and claiming the technological frontier.

We believe in the romance of technology, of industry. The eros of the train, the car, the electric light, the skyscraper. And the microchip, the neural network, the rocket, the split atom.

We believe in adventure. Undertaking the Hero's Journey, rebelling against the status quo, mapping uncharted territory, conquering dragons, and bringing home the spoils for our community.

To paraphrase a manifesto of a different time and place:

> Beauty exists only in struggle. There is no masterpiece that has not an aggressive character. Technology must be a violent assault on the forces of the unknown, to force them to bow before man.

We believe that we are, have been, and will always be the masters of technology, not mastered by technology. Victim mentality is a curse in every domain of life, including in our relationship with technology—both unnecessary and self-defeating. We are not victims, we are conquerors.

We believe in nature, but we also believe in overcoming nature. We are not primitives, cowering in fear of the lightning bolt. We are the apex predator; the lightning works for us.

We believe in greatness. We admire the great technologists and industrialists who came before us, and we aspire to make them proud of us today.

And we believe in humanity—individually and collectively.

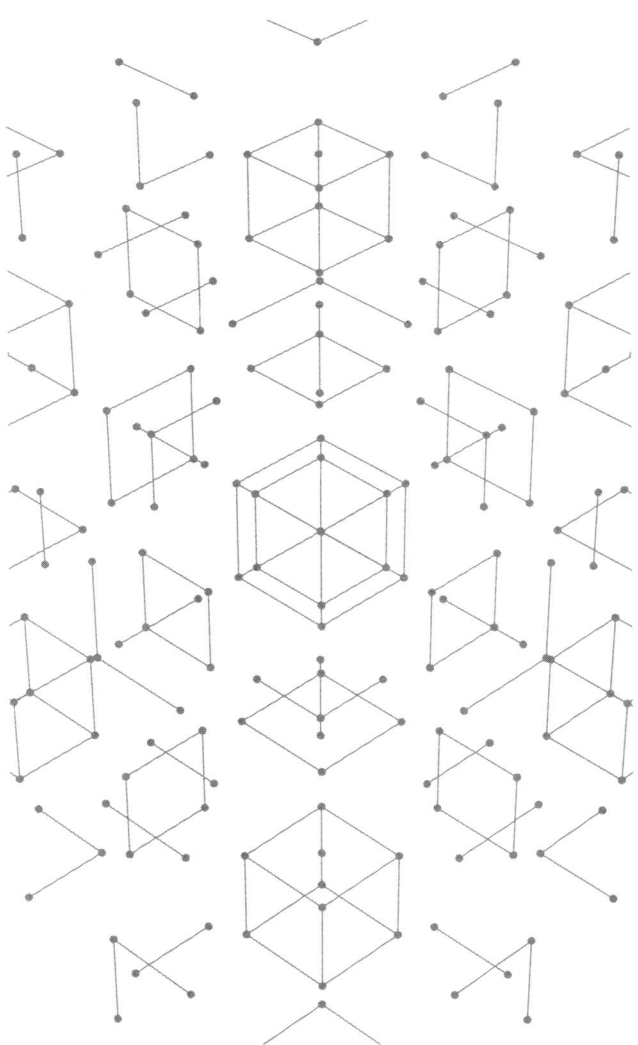

Technological Values

We believe in ambition, aggression, persistence, relentlessness—strength.

We believe in merit and achievement.

We believe in bravery, in courage.

We believe in pride, confidence, and self-respect—when earned.

We believe in free thought, free speech, and free inquiry.

We believe in the actual Scientific Method and Enlightenment values of free discourse and challenging the authority of experts.

We believe, as Richard Feynman said,

> Science is the belief in the ignorance of experts.

And,

> I would rather have questions that can't be answered than answers that can't be questioned.

We believe in local knowledge, the people with actual information making decisions, not in playing God.

We believe in embracing variance, in increasing interestingness.

We believe in risk, in leaps into the unknown.

We believe in agency, in individualism.

We believe in radical competence.

We believe in an absolute rejection of resentment. As Carrie Fisher said, "Resentment is like drinking poison and waiting for the other person to die." We take responsibility and we overcome.

We believe in competition, because we believe in evolution.

We believe in evolution, because we believe in life.

We believe in the truth.

We believe rich is better than poor, cheap is better than expensive, and abundant is better than scarce.

We believe in making everyone rich, everything cheap, and everything abundant.

We believe extrinsic motivations—wealth, fame, revenge—are fine as far as they go. But we believe intrinsic motivations—the satisfaction of building something new, the camaraderie of being on a team, the achievement of becoming a better version of oneself—are more fulfilling and more lasting.

We believe in what the Greeks called *eudaimonia* through *arete*—flourishing through excellence.

We believe technology is universalist. Technology doesn't care about your ethnicity, race, religion, national origin, gender, sexuality, political views, height, weight, hair or lack thereof. Technology is built by a virtual United Nations of talent from all over the world. Anyone with a positive attitude and a cheap laptop can contribute. Technology is the ultimate open society.

We believe in the Silicon Valley code of "pay it forward," trust via aligned incentives, generosity of spirit to help one another learn and grow.

We believe America and her allies should be strong and not weak. We believe national strength of liberal democracies flows from economic strength (financial power), cultural strength (soft power), and military strength (hard power). Economic, cultural, and military strength flow from technological strength. A technologically strong America is a force for good in a dangerous world. Technologically strong liberal democracies safeguard liberty and peace. Technologically weak liberal democracies lose to their autocratic rivals, making everyone worse off.

We believe technology makes greatness more possible and more likely.

We believe in fulfilling our potential, becoming fully human—for ourselves, our communities, and our society.

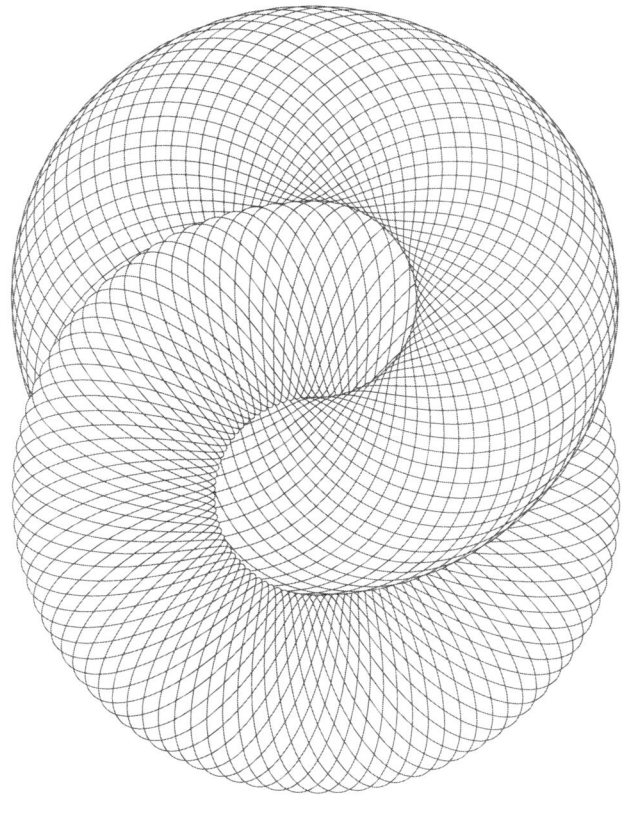

The Meaning of Life

Techno-Optimism is a material philosophy, not a political philosophy.

We are not necessarily left wing, although some of us are.

We are not necessarily right wing, although some of us are.

We are materially focused, for a reason—to open the aperture on how we may choose to live amid material abundance.

A common critique of technology is that it removes choice from our lives as machines make decisions for us. This is undoubtedly true, yet it is more than offset by the freedom to create our lives that flows from the material abundance created by our use of machines.

Material abundance from markets and technology opens the space for religion, for politics, and for choices of how to live, socially and individually.

We believe technology is liberatory. Liberatory of human potential. Liberatory of the human soul, the human spirit. Expanding what it can mean to be free, to be fulfilled, to be alive.

We believe technology opens the space of what it can mean to be human.

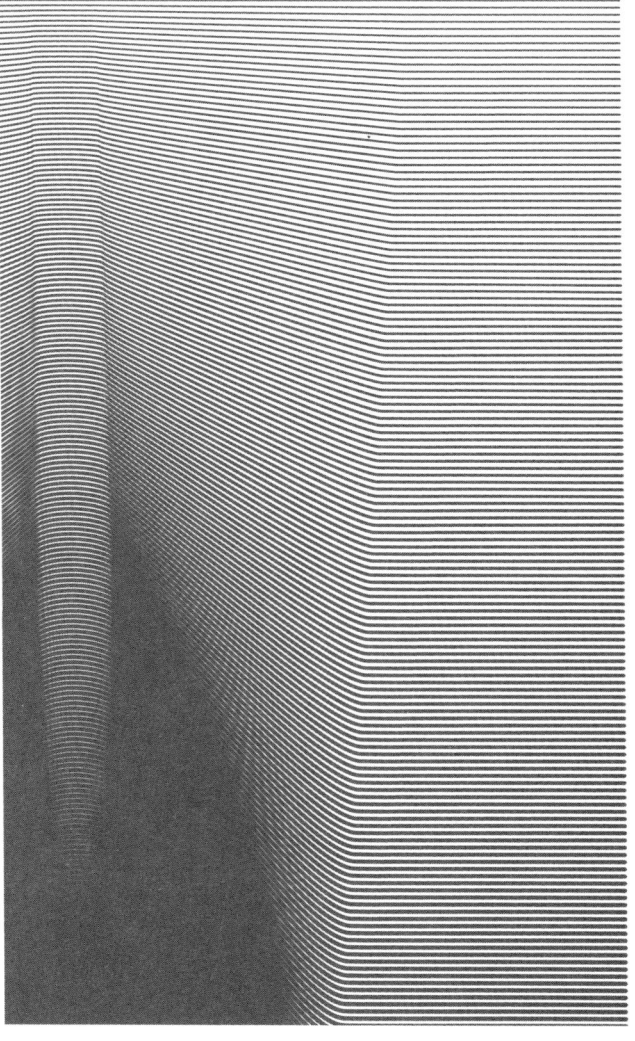

The Enemy

We have enemies.

Our enemies are not bad people—but rather bad ideas.

Our present society has been subjected to a mass demoralization campaign for six decades—against technology and against life—under different names such as "existential risk," "sustainability," "ESG," "Sustainable Development Goals," "social responsibility," "stakeholder capitalism," "Precautionary Principle," "trust and safety," "tech ethics," "risk management," "de-growth," "the limits of growth."

This demoralization campaign is based on bad ideas of the past—zombie ideas, many derived from Communism, disastrous then and now—that have refused to die.

Our enemy is stagnation.

Our enemy is anti-merit, anti-ambition, anti-striving, anti-achievement, anti-greatness.

Our enemy is statism, authoritarianism, collectivism, central planning, socialism.

Our enemy is bureaucracy, vetocracy, gerontocracy, blind deference to tradition.

Our enemy is corruption, regulatory capture, monopolies, cartels.

Our enemy is institutions that in their youth were vital and energetic and truth-seeking, but are now compromised and corroded and collapsing—blocking progress in increasingly desperate bids for continued relevance, frantically trying to justify their ongoing funding despite spiraling dysfunction and escalating ineptness.

Our enemy is the ivory tower, the know-it-all credentialed expert worldview, indulging in abstract theories, luxury beliefs, social engineering, disconnected from the real world, delusional, unelected, and unaccountable—playing God with everyone else's lives, with total insulation from the consequences.

Our enemy is speech control and thought control—the increasing use, in plain sight, of George Orwell's *1984* as an instruction manual.

Our enemy is Thomas Sowell's Unconstrained Vision, Alexandre Kojève's Universal and Homogeneous State, Thomas More's Utopia.

Our enemy is the Precautionary Principle, which would have prevented virtually all progress since man first harnessed fire. The Precautionary Principle was invented to prevent the large-scale deployment of civilian nuclear power, perhaps the most catastrophic mistake in Western society in my lifetime. The Precautionary Principle continues to inflict enormous unnecessary suffering on our world today. It is deeply immoral, and we must jettison it with extreme prejudice.

Our enemy is deceleration, degrowth, depopulation—the nihilistic wish, so trendy among our elites, for fewer people, less energy, and more suffering and death.

Our enemy is Friedrich Nietzsche's Last Man:

> I tell you: one must still have chaos in oneself, to give birth to a dancing star. I tell you: you have still chaos in yourselves.
>
> Alas! There comes the time when man will no longer give birth to any star. Alas! There comes the time of the most despicable man, who can no longer despise himself …
>
> "What is love? What is creation? What is longing?

What is a star?"—so asks the Last Man, and blinks.

The earth has become small, and on it hops the Last Man, who makes everything small. His species is ineradicable as the flea; the Last Man lives longest …

One still works, for work is a pastime. But one is careful lest the pastime should hurt one.

One no longer becomes poor or rich; both are too burdensome …

No shepherd, and one herd! Everyone wants the same; everyone is the same: he who feels differently goes voluntarily into the madhouse.

"Formerly all the world was insane,"—say the subtlest of them, and they blink.

They are clever and know all that has happened: so there is no end to their derision …

"We have discovered happiness,"—say the Last Men, and they blink.

Our enemy is … that.

We aspire to be ... not that.

We will explain to people captured by these zombie ideas that their fears are unwarranted and the future is bright.

We believe these captured people are suffering from ressentiment—a witches' brew of resentment, bitterness, and rage that is causing them to hold mistaken values, values that are damaging to both themselves and the people they care about.

We believe we must help them find their way out of their self-imposed labyrinth of pain.

We invite everyone to join us in Techno-Optimism.

The water is warm.

Become our allies in the pursuit of technology, abundance, and life.

The Future

Where did we come from?

Our civilization was built on a spirit of discovery, of exploration, of industrialization.

Where are we going?

What world are we building for our children and their children, and their children?

A world of fear, guilt, and resentment?

Or a world of ambition, abundance, and adventure?

We believe in the words of David Deutsch:

> We have a duty to be optimistic. Because the future is open, not predetermined and therefore cannot just be accepted: we are all responsible for what it holds. Thus it is our duty to fight for a better world.

We owe the past, and the future.

It's time to be a Techno-Optimist.

It's time to build.

MARC ANDREESSEN

Patron Saints of Techno-Optimism

In lieu of detailed endnotes and citations, read the work of these people, and you too will become a Techno-Optimist.

@BasedBeffJezos
@bayeslord
@PessimistsArc
Ada Lovelace
Adam Smith
Andy Warhol
Bertrand Russell
Brad DeLong
Buckminster Fuller
Calestous Juma

Clayton Christensen
Dambisa Moyo
David Deutsch
David Friedman
David Ricardo
Deirdre McCloskey
Doug Engelbart
Elting Morison
Filippo Tommaso Marinetti
Frederic Bastiat
Frederick Jackson Turner
Friedrich Hayek
Friedrich Nietzsche
George Gilder
Isabel Paterson
Israel Kirzner
James Burnham
James Carse
Joel Mokyr
Johan Norberg
John Galt
John Von Neumann
Joseph Schumpeter

Julian Simon
Kevin Kelly
Louis Rossetto
Ludwig von Mises
Marian Tupy
Martin Gurri
Matt Ridley
Milton Friedman
Neven Sesardić
Nick Land
Paul Collier
Paul Johnson
Paul Romer
Ray Kurzweil
Richard Feynman
Rose Wilder Lane
Stephen Wolfram
Stewart Brand
Thomas Sowell
Vilfredo Pareto
Virginia Postrel
William Lewis
William Nordhaus

Marc Andreessen is a co-founder and general partner at the venture capital firm Andreessen Horowitz. He is an innovator and creator, one of the few to pioneer a software category used by more than a billion people and one of the few to establish multiple billion-dollar companies.

Marc co-created the highly influential Mosaic internet browser and co-founded Netscape, which later sold to AOL for $4.2 billion. He also co-founded Loudcloud, which as Opsware, sold to Hewlett-Packard for $1.6 billion. He later served on the board of Hewlett-Packard from 2008 to 2018.

Marc holds a BS in computer science from the University of Illinois at Urbana-Champaign.

Marc serves on the board of the following Andreessen Horowitz portfolio companies: Applied Intuition, Carta, Coinbase, Dialpad, Flow, Golden, Honor, OpenGov, Samsara, Simple Things, and TipTop Labs. He is also on the board of Meta.